How to Sparkle at
Number Bonds

Jean Haigh
Beryl Webber

Brilliant
PUBLICATIONS

We hope you and your class enjoy using this book. To find out more details on any of the other books in the series listed below, please log onto our website: www.brilliantpublications.co.uk.

Maths titles
How to Sparkle at Addition and Subtraction to 20   978 1 897675 28 1
How to Sparkle at Beginning Multiplication and Division   978 1 897675 30 4
How to Sparkle at Maths Fun   978 1 897675 86 1
How to Sparkle at Counting to 10   978 1 897675 27 4

Science titles
How to Sparkle at Assessing Science   978 1 897675 20 5
How to Sparkle at Science Investigations   978 1 897675 36 6

English titles
How to Sparkle at Alphabet Skills   978 1 897675 17 5
How to Sparkle at Grammar and Punctuation   978 1 897675 19 9
How to Sparkle at Nursery Rhymes   978 1 897675 16 8
How to Sparkle at Phonics   978 1 897675 14 4
How to Sparkle at Prediction Skills   978 1 897675 15 1
How to Sparkle at Word Level Activities   978 1 897675 90 8
How to Sparkle at Writing Stories and Poems   978 1 897675 18 2
How to Sparkle at Reading Comprehension   978 1 903853 44 3

Festive title
How to Sparkle at Christmas Time   978 1 897675 62 5

Published by Brilliant Publications
Unit 10
Sparrow Hall Farm
Edlesborough
Dunstable
Bedfordshire
LU6 2ES, UK

email: info@brilliantpublications.co.uk

General information enquiries:
Tel: 01525 222292

The name Brilliant Publications and the logo are registered trademarks.

Written by Jean Haigh and Beryl Webber
Illustrated by Kate Ford

Printed in the UK.

© Jean Haigh and Beryl Webber 1998

Printed ISBN: 978-1-897675-34-2
ebook ISBN: 978-0-85747-062-1

First published 1998. Reprinted 2001 and 2009.
10 9 8 7 6 5 4 3

Contents

Introduction

How to Sparkle at Number Bonds contains 42 photocopiable ideas for use with 5 – 7 year olds. The book provides a flexible, but structured resource for teaching children to understand and use number bonds. The ultimate aim is for children to be able to calculate at speed by heart. The book does not cover the complete programme of study. The children will need fingers and counters until they have begun to internalize the number bonds. Some activities require the use of coloured pencils.

The activities can be used with individual children or small groups as the need arises. The text on each page has been kept as short as possible so that the children will feel confident to tackle the sheets without too much teacher input. Some children may require you to read and talk through the page carefully and to discuss the mathematics with them.

The activities should be used flexibly and can be used whenever the need arises for particular activities to support and supplement your core mathematics programme. It is not intended that every child should complete every activity, nor is this book intended to provide a complete mathematics programme. The order in which the pages are arranged is not necessarily hierarchical and children may complete them in any order.

The last page of the book is a record sheet for the children to keep track of which activities they have undertaken.

How to Sparkle at Number Bonds relates directly to the programmes of study for Using and Applying Mathematics and Number at Key Stage 1. The page opposite gives further details. On the contents page the activities are coded according to the level of difficulty with a letter code (A–C). This is provided to give you an indication of how the activities relate to mathematical progression within the key stage. Activities coded A are the most challenging in terms of this book.

Links to the National Curriculum

The activities in this book allow children to have opportunities to:

Ma2 NUMBER
Using and applying number
1   Pupils should be taught to:
Problem solving
a   approach problems involving number, and data presented in a variety of forms, in order to identify what they need to do

b   develop flexible approaches to problem solving and look for ways to overcome difficulties

c   make decisions about which operations and problem-solving strategies to use

d   organize and check their work
Communicating
e.   use the correct language, symbols and vocabulary associated with number and data

f   communicate in spoken, pictorial and written form, at first using informal language and recording, then mathematical language and symbols
Reasoning
g   present results in an organized way

i   explain their methods and reasoning when solving problems involving number and data.

Numbers and the number system
2   Pupils should be taught to:
Counting
a   count reliably up to 20 objects; be familiar with the numbers 11 to 20; gradually extend counting to 100 and beyond
Number patterns and sequences
b   create and describe number patterns; explore and record patterns related to addition and subtraction and then patterns of multiples of 2, 5 and 10 explaining the patterns and using them to make predictions.
The number system
c   read and write numbers to 20 at first and then to 100 or beyond; understand and use the vocabulary of comparing and ordering these numbers; recognize that the position of a digit gives its value and know what each digit represents

Calculations
3   Pupils should be taught to:
Number operations and the relationship between them
a   understand addition and use related vocabulary; recognize that addition can be done in any order; understand subtraction as both 'take away' and 'difference' and use the related vocabulary; recognize that subtraction is the inverse of addition; use the symbol '=' to represent equality; solve simple missing number problems

4   Pupils should be taught to:
Solving numerical problems
a   choose sensible calculation methods to solve whole-number problems, drawing on their understanding of the operations.

Cakes

Ram wants to know how many cakes are left.

He notices there are four cakes and three cakes.

$$4 + 3 = 7$$

How many cakes are on these plates? Write the sums.

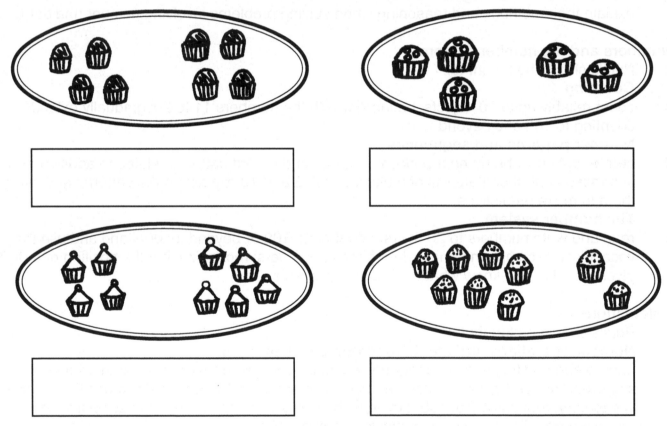

Colour the star if you can pick up a handful of cubes and know quickly how many cubes there are.

Cherries

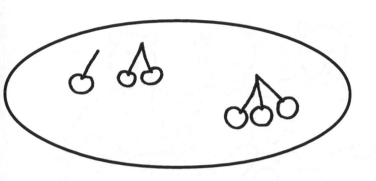

Count the cherries.
Write the sum.

| 1 | + | 2 | + | 3 | = | 6 |

Now try these:

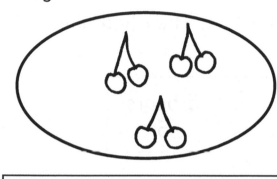

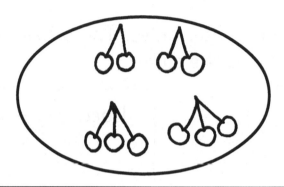

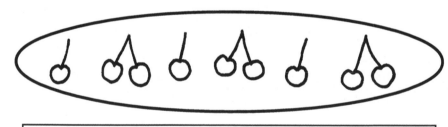

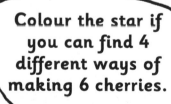

Colour the star if you can find 4 different ways of making 6 cherries.

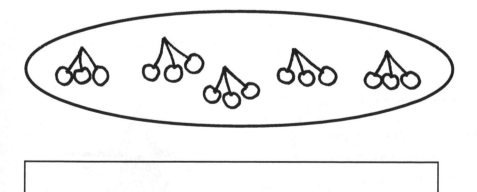

Teddy bears

One teddy hides. The other bears look for him. They find him.
This is the sum:

1 bear	+	4 bears	=	5 bears
1	+	4	=	5

Make up a teddy story of your own and write the sum.

Colour the star if you
can find all the sums
that make 5.

Ssssnakes

Finish writing numbers in the snake.

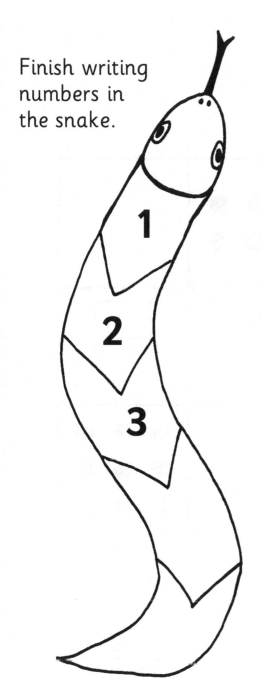

1

2

3

The snake turns round. Write the numbers.

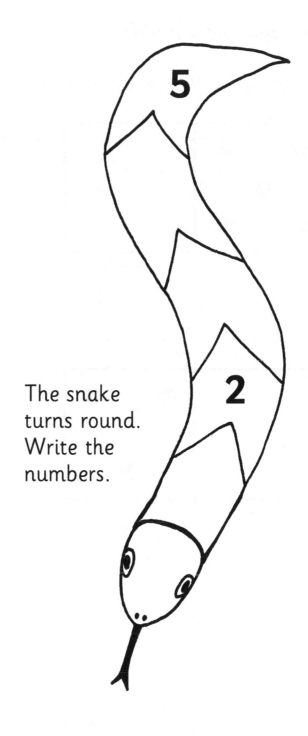

5

2

Colour the star if you can count backwards from 5 to 1.

Dominoes

All the dominoes total 10.

Put in the spots. Write the numbers.

5 [ ]

6 [ ]

[ ] 3

[ ] [ ]

[ ] [ ]

Colour the star if you can find the two dominoes that are the same.

Numbers

Fill in the missing numbers.

| 1 | 2 | 3 | | | | | | | |

| | | | | 5 | 6 | 7 | | | |

| | | 11 | 12 | 13 | | | | | |

| | | | | | | | 18 | 19 | 20 |

Colour the star if you can count to 50.

Number order, 1

1	2	3	4	5	6	7	8	9	10

Put a counter on 3.   What is 1 more?

3   +   1   =   △

What is 2 more?

3   +   2   =   △

What is 3 more?

3   +   3   =   △

Make up some sums of your own.

Start on 5.   What is ◯ more?

5   +   ◯   =   △

Start on ☐ .   What is ◯ more?

☐   +   ◯   =   △

Colour the star if you know what is 3 more than 10.

Number order, 2

1	2	3	4	5	6	7	8	9	10	11	12	13	14	15	16	17	18	19	20

Write the number that is one more:

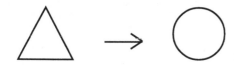

Write the number that is ten more:

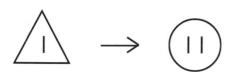

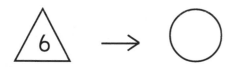

Colour the star if you know what number is ten more than 20.

Make 5 ducks

$1 \text{ duck} + 4 \text{ ducks} = 5 \text{ ducks}$

$1 + 4 = 5$

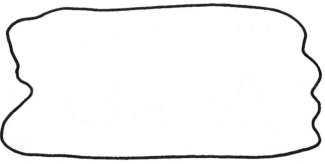

2 ducks + 3 ducks = 5 ducks

| | + | | = | 5 |

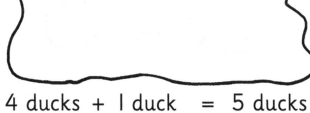

4 ducks + 1 duck = 5 ducks

| | + | | = | 5 |

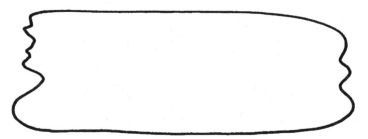

3 ducks + 2 ducks = 5 ducks

| | + | | = | 5 |

5 ducks + 0 ducks = 5 ducks

| | + | | = | |

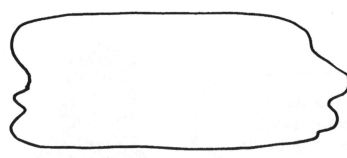

0 ducks + 5 ducks = 5 ducks

| | + | | = | |

Colour the star if you can make up a number story about the ducks.

Socks

Count the socks. ☐

Draw lines to match the pairs. Count the pairs. ☐

Colour the star if you can count in twos to 20.

Dragon story

Choose 4 dragons. Put a counter on each one.

How many are left?

$$10 - 4 = 6$$

Do the same for other numbers of dragons. Write the sums.

Colour the star if you found all the sums you can.

Ten sums

Finish the pattern.

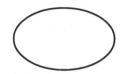 + 1 = 10

2 +  = 10

 + 3 =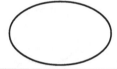

4 +  =

 + 5 =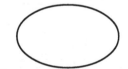

6 +  =

 + 7 =

8 + 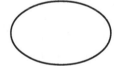 =

 + 9 =

10 +  =

Colour the star if you can remember these sums by heart.

Ten sums again

Make sums that total 10.

◯ + △ + ☐ = 10

◯ + △ + ☐ = 10

◯ + △ + ☐ = 10

◯ + ◯ + ☐ = 10

◯ + △ + △ = 10

☐ + △ + ☐ = 10

◯ + △ + ☐ + ⬡ = 10

◯ + △ + ☐ + ⬡ = 10

◯ + △ + ☐ + ⬡ = 10

Colour the star if you can find
5 numbers that total 10.

All about ten

How quickly can you do these?

1 + 9 = ☐   10 – 6 = ☐

5 + ☐ = 10   10 – ☐ = 8

3 + 7 = ☐   8 + ☐ = 10

10 – 3 = ☐   ☐ + 7 = 10

0 + ☐ = 10   10 – 9 = ☐

Do these sums. What do you notice about them?

1 + 9 = ☐   9 + ☐ = 10

2 + ☐ = 10   ☐ + 2 = 10

☐ + 7 = ☐   7 + 3 = ☐

☐ + 6 = ☐   6 + ☐ = ☐

☐ + ☐ = 10   ☐ + ☐ = 10

Colour the star
if you know all
the sums to 10.

Colour chart

Here is a chart of Class 3's favourite colours.

Colour	Number of people
Blue	6
Red	9
Yellow	5
Green	3
Orange	1
White	4
Pink	2

Which colour was the most popular?

Which colour was the least popular?

How many children liked yellow and orange?

How many children liked red and pink?

How many people altogether?

Colour the star if you can order the colours from the most to the least popular.

Apples

Two children share the apples.
They have [   ] apples each.

Three children share the apples.
How many do they have each?

What about four children?

What about six children?

Colour the star if you can share 20 apples between 4 children.

Add 9

2 —— + 10 ——→ 12 —— − 1 ——→ 11

6 ——→ 16 ——→ ◯

9 ——→ ◯ ——→ ◯

5 ——→ ◯ ——→ ◯

7 ——→ ◯ ——→ ◯

3 ——→ ◯ ——→ ◯

4 ——→ ◯ ——→ ◯

What do you notice about these numbers?

Colour the star if you can always add 9 in your head.

Star sums

The addition sums in the star all make the number in the middle.
Finish the stars.

Colour the star if you can find all the addition sums that make 12.

Star subtracts

The number in the middle of the star is the answer to all the take away sums. Finish the stars.

Colour the star if you can tell your teacher about all the take aways with the answer 12.

Quick sums

Try these sums with a friend. Who finishes first? Were all the answers correct?

$20 - 7 =$	$20 - 9 =$
$17 - 6 =$	$18 + 5 =$
$25 - 4 =$	$26 - 2 =$
$14 + 9 =$	$16 + 7 =$
$4 + 2 + 1 =$	$3 + 1 + 2 =$
$10 + 10 + 10 =$	$10 + 10 =$
$11 + 3 - 1 =$	$12 + 4 - 2 =$
$19 - 12 =$	$18 - 11 =$
$100 - 50 =$	$100 - 40 =$
$100 + 20 + 6 =$	$100 + 30 + 4 =$

Colour the star if you can do all of these sums in your head.

How much money?

Count how much money is in each purse.

Colour the star if you can find 5 coins to make £1.

Pencil case

You have £1 to spend. What can you buy?

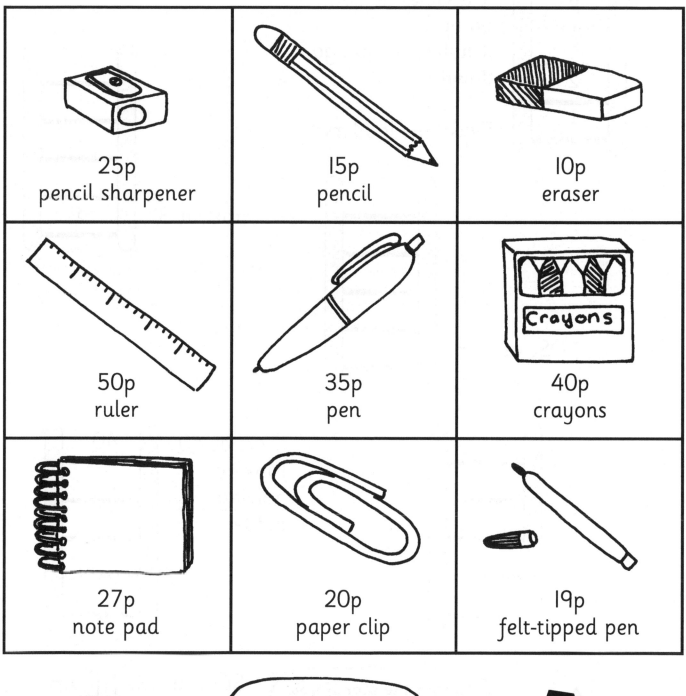

25p pencil sharpener	15p pencil	10p eraser
50p ruler	35p pen	40p crayons
27p note pad	20p paper clip	19p felt-tipped pen

Colour the star if you can find 3 things that cost 50p all together.

Ladders

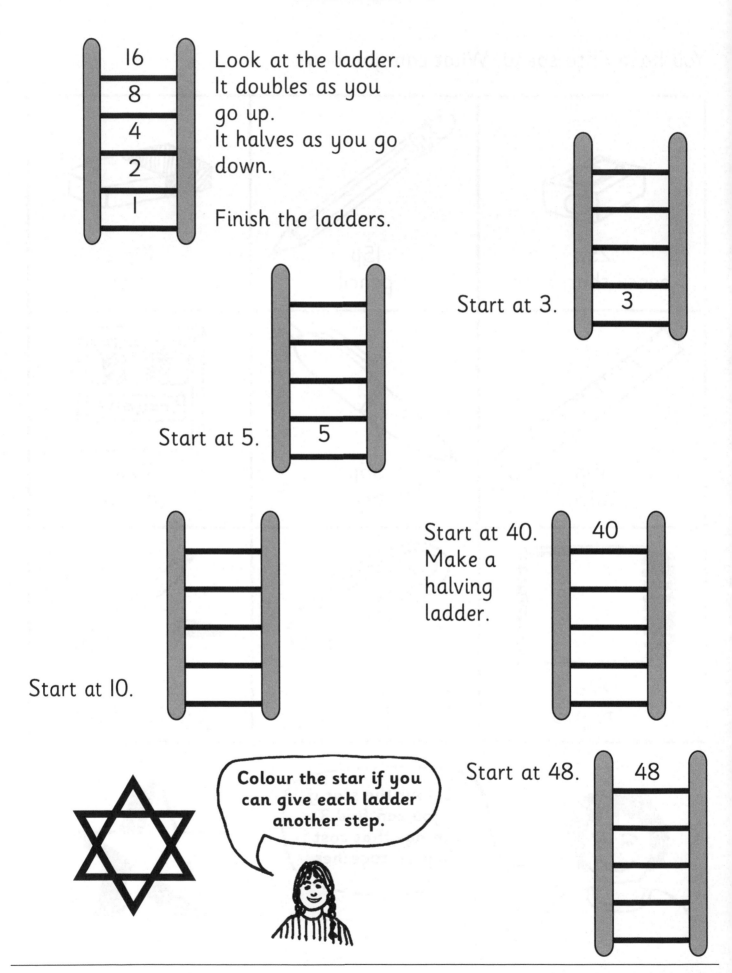

16
8
4
2
1

Look at the ladder.
It doubles as you
go up.
It halves as you go
down.

Finish the ladders.

Start at 3.   3

Start at 5.   5

Start at 10.

Start at 40.   40
Make a
halving
ladder.

Colour the star if you
can give each ladder
another step.

Start at 48.   48

Even numbers

Fill in the missing numbers.

| 2 | 4 | 6 | | | | | | | |

| | | 6 | 8 | 10 | | | | | |

| 14 | 16 | 18 | | | | | | | |

| | | | | | | | 46 | 48 | 50 |

Colour the star if you can count in twos to 50 and 100.

Odd and even balloons

Colour the balloons red if the number is even.
Colour the balloons yellow if the number is odd.

Colour the star
if you have 8
yellow balloons.

Make six

Find different ways of making 6 using 3 numbers.
Use numbers 1, 2, 3, 4. You can use them more than once.

1	+	1	+	4	=	6
	+		+		=	6
	+		+		=	6
	+		+		=	6
	+		+		=	6
	+		+		=	6
	+		+		=	6
	+		+		=	6
	+		+		=	6
	+		+		=	6

Check the sums are all different. If they are, you have found all the different ways of making 6 by adding three of these numbers.

Colour the star if can can find all the ways of making 6 using 1, 2, 3, 5. There are another 10 ways.

Two steps to twenty

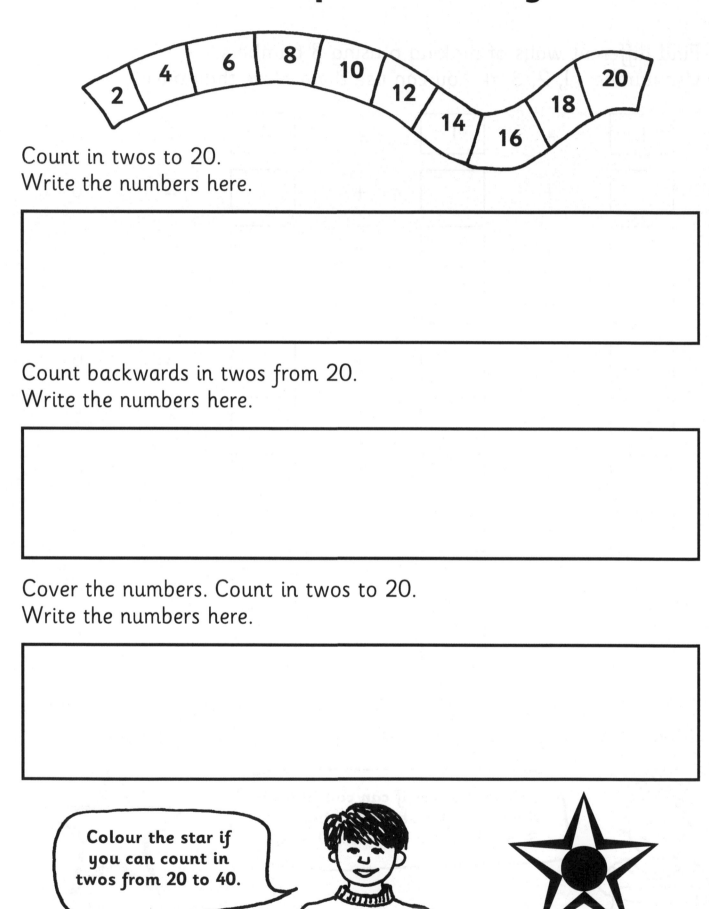

Count in twos to 20.
Write the numbers here.

Count backwards in twos from 20.
Write the numbers here.

Cover the numbers. Count in twos to 20.
Write the numbers here.

Colour the star if you can count in twos from 20 to 40.

Five steps to fifty

5	10	15	20	25	30	35	40	45	50

Count in fives to 50.
Write the numbers here.

Count backwards in fives from 50.
Write the numbers here.

Cover the numbers. Count in fives to 50.
Write the numbers here.

Colour the star if you can count in fives from 50 to 100.

Stamps

Jon and Jane buy some stamps.

Jon buys two 10p stamps.

The stamps cost 2 x 10p ——→ 20p.

Jane buys three 5p stamps.

Write the sum here.

How much do they cost?

Jon buys four 2p stamps.

Write the sum here.

How much do they cost?

Jon buys five 10p stamps.

Write the sum here.

How much do they cost?

Jane buys four 5p stamps.

Write the sum here.

How much do they cost?

Colour the star if you can make up four more stamp sums.

How many can I do?

Colour the apple red if you can work out the sum in your head.

Colour the apple green if you can work out the sum using fingers or counters.

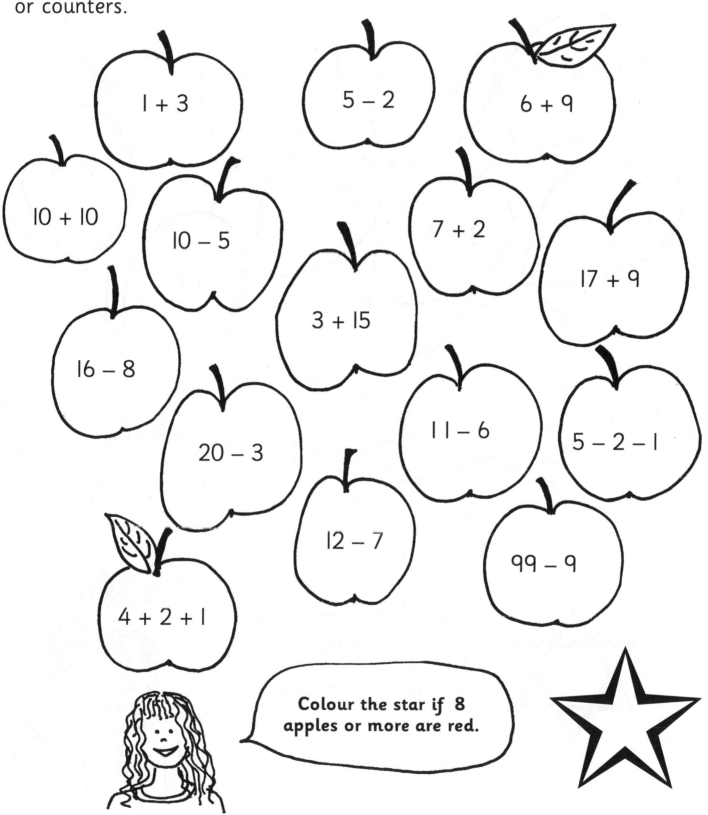

1 + 3

5 − 2

6 + 9

10 + 10

10 − 5

7 + 2

17 + 9

3 + 15

16 − 8

20 − 3

11 − 6

5 − 2 − 1

12 − 7

99 − 9

4 + 2 + 1

Colour the star if 8 apples or more are red.

Balloons

Colour the balloon red if the sum makes 10.
Colour the balloon blue if the sum makes 9.
Colour the balloon yellow if the sum makes 11.

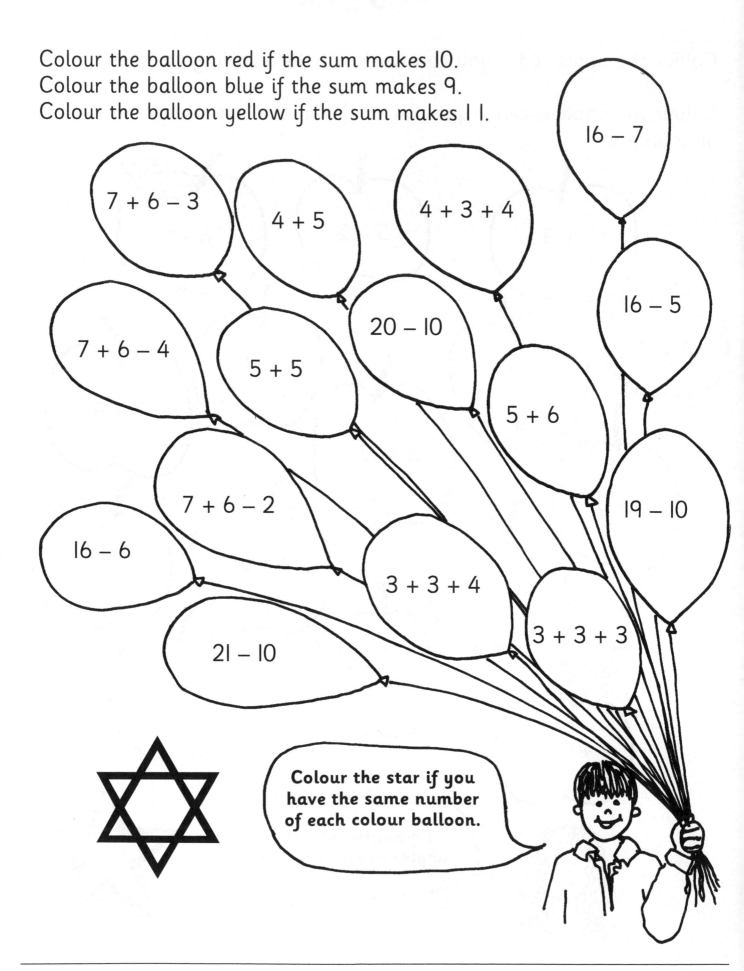

16 – 7

7 + 6 – 3

4 + 5

4 + 3 + 4

16 – 5

7 + 6 – 4

5 + 5

20 – 10

5 + 6

19 – 10

16 – 6

7 + 6 – 2

3 + 3 + 4

3 + 3 + 3

21 – 10

Colour the star if you have the same number of each colour balloon.

The bus, 1

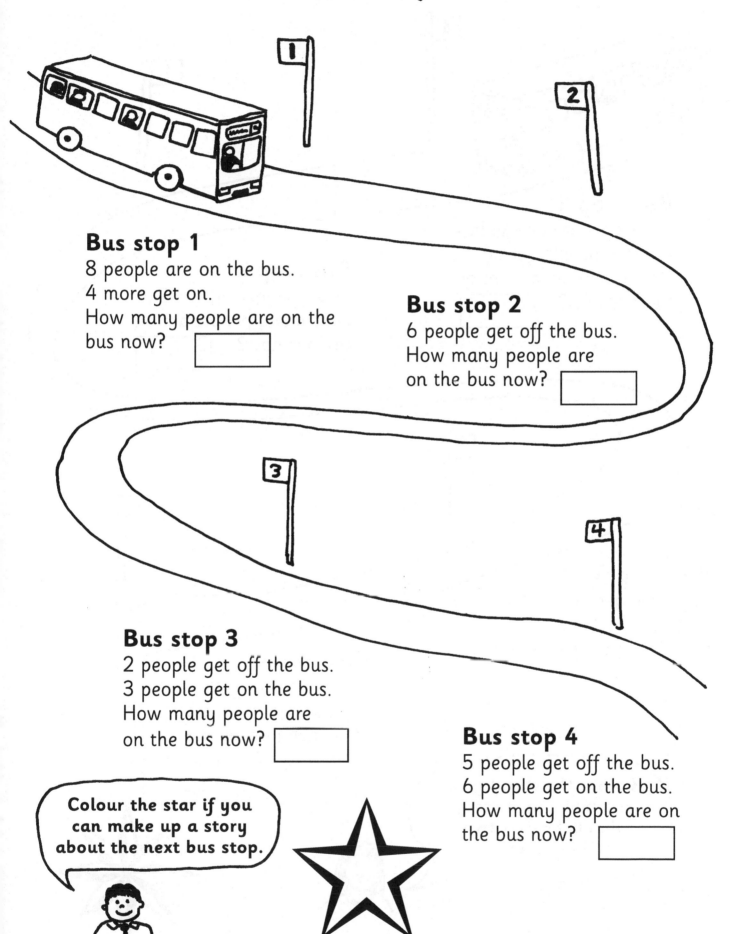

Bus stop 1

8 people are on the bus.
4 more get on.
How many people are on the
bus now?

Bus stop 2

6 people get off the bus.
How many people are
on the bus now?

Bus stop 3

2 people get off the bus.
3 people get on the bus.
How many people are
on the bus now?

Bus stop 4

5 people get off the bus.
6 people get on the bus.
How many people are on
the bus now?

Colour the star if you
can make up a story
about the next bus stop.

The bus, 2

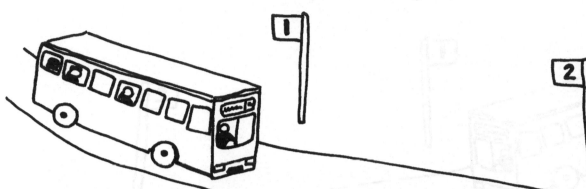

Bus stop 1

20 people are on the bus.
14 more people get on.
How many people are on
the bus now?

Bus stop 2

13 people get off the bus.
How many people are on
the bus now?

Bus stop 3

11 people get off the bus.
23 people get on.
How many people are on
the bus now?

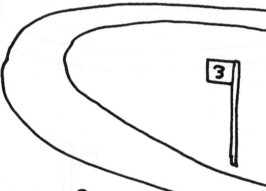

Bus stop 4

17 people get off the bus.
4 people get on the bus.
How many people are on
the bus now?

Colour the star if you
can make up a story
about the next two
bus stops.

Number square

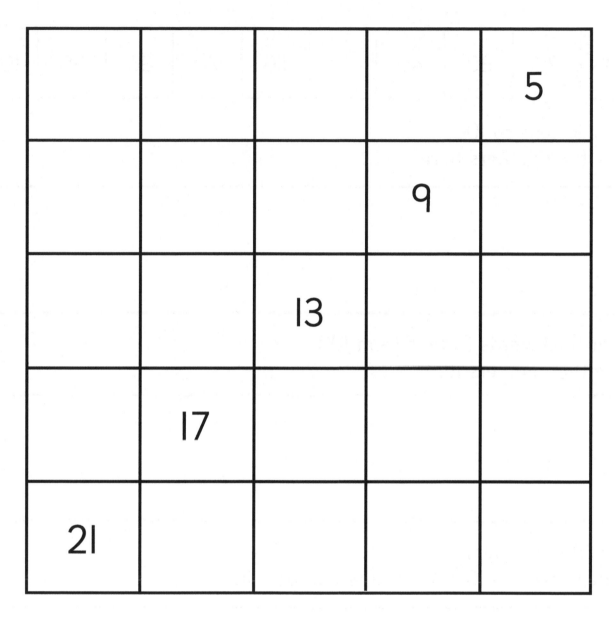

				5
			9	
		13		
	17			
21				

Complete the grid.

What do you notice about the numbers in the corners?

What do you notice about the even numbers?

Colour the star if you notice anything else about the grid.

One hundred

10	20	30	40	50	60	70	80	90	100

Count in tens to 100.
Write the numbers here.

Count backwards in tens from 100.
Write the numbers here.

Cover the numbers.
Count in tens to 100. Write the numbers here.

Colour the star if you can count backwards in tens from 100 without looking at the numbers.

Ten more

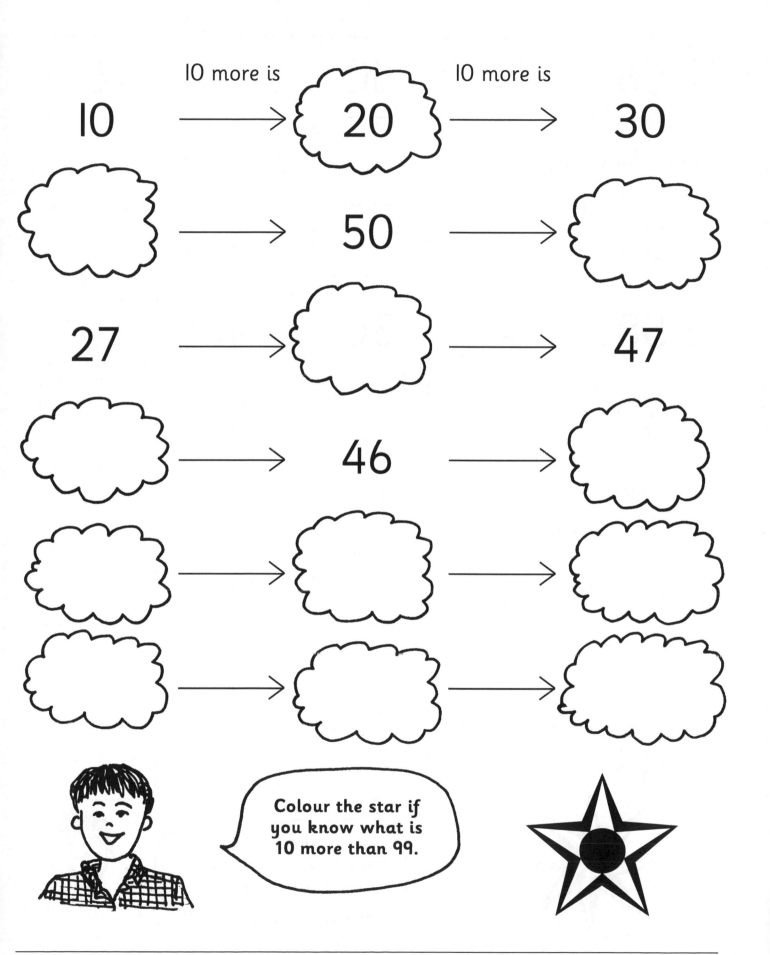

10 more is

10 more is

10 ⟶ 20 ⟶ 30

⟶ 50 ⟶

27 ⟶ ⟶ 47

⟶ 46 ⟶

Colour the star if you know what is 10 more than 99.

Tens and units

| 56 | = | 50 | + | 6 |

Finish the sums.

54 = [ ] + 4

58 = 50 + [ ]

32 = [ ] + [ ]

17 = 10 + [ ]

86 = [ ] + [ ]

47 = [ ] + 7

99 = [ ] + [ ]

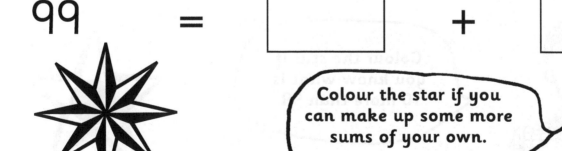

Colour the star if you
can make up some more
sums of your own.

Two hundred

| 10 | 20 | 30 | 40 | 50 | 60 | 70 | 80 | 90 | 100 | 110 | 120 | 130 | 140 | 150 | 160 | 170 | 180 | 190 | 200 |

Count in tens to 200.

Count backwards in tens from 200.

Count in twenties to 200.

Count backwards in twenties from 200.

Write the number patterns here.

Colour the star if you can count in tens to 1000.

Missing numbers

Here is part of a 100 square.
Fill in the missing numbers.

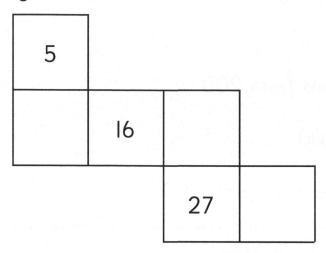

Fill in the missing numbers.

Make up one of your own.

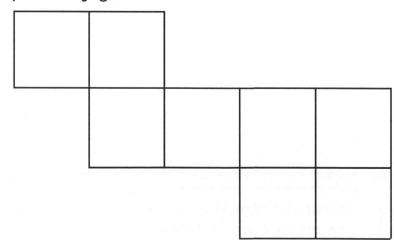

Colour the star if you can find where these fit on a 100 square.

Think of a number, 1

I think of a number
and add 1.
My answer is 4.
What is my number?

I think of a number
and take away 2.
My answer is 7.
What is my number?

I think of a number and
add 3 then take away 1.
My answer is 5.
What is my number?

I think of a number and
add 5 then add 3.
My answer is 10.
What is my number?

I think of a number
and add 2.
My answer is 7.
What is my number?

Colour the star if you can
make up another number
puzzle of your own.

Think of a number, 2

Our answers are always 10. Can you find the numbers we are thinking of?

I think of this number and add 6. What is my number?

I think of a number and take away 2. What is my number?

I think of a number and add 8 then take away 5. What is my number?

I think of a number and add 5 then take away 3. What is my number?

I think of a number and add 6 then take away 6. What is my number?

I think of a number and take away 7 then add 5. What is my number?

I think of a number and add 5 then add 5. What is my number?

I think of a number and take away 5 then take away 5. What is my number?

Colour the star if you can make up another number puzzle where the answer is 10.

Think of a number, 3

I think of a number and double it. The answer is 4. What is my number?

I think of a number and double it. The answer is 10. What is my number?

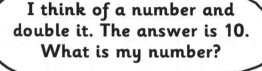

I think of a number and double it. The answer is 6. What is my number?

I think of a number and double it and add 1 then take away 2. The answer is 7. What is my number?

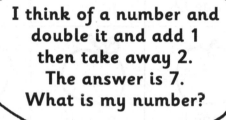

I think of a number and double it and add 2. The number is 6. What is my number?

I think of a number and double it and add 1. The answer is 5. What is my number?

I think of a number and double it and take away 1. The answer is 5. What is my number?

Colour the star if you can make up a doubling number puzzle.

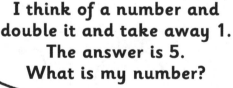

I have ...

Colour the apple if you have finished the page.

page 6	page 7	page 8	page 9	page 10	page 11
page 12	page 13	page 14	page 15	page 16	page 17
page 18	page 19	page 20	page 21	page 22	page 23
page 24	page 25	page 26	page 27	page 28	page 29
page 30	page 31	page 32	page 33	page 34	page 35
page 36	page 37	page 38	page 39	page 40	page 41
page 42	page 43	page 44	page 45	page 46	page 47